Zofia Kiliańska
Jolanta D. Żołnierczyk
Arleta Borowiak

Responses of CLL cells to purine analogs with cyclophosphamide

AF153263

Zofia Kiliańska
Jolanta D. Żołnierczyk
Arleta Borowiak

Responses of CLL cells to purine analogs with cyclophosphamide

Drug-induced apoptosis of leukemic cells

LAP LAMBERT Academic Publishing

Impressum / Imprint

Bibliografische Information der Deutschen Nationalbibliothek: Die Deutsche Nationalbibliothek verzeichnet diese Publikation in der Deutschen Nationalbibliografie; detaillierte bibliografische Daten sind im Internet über http://dnb.d-nb.de abrufbar.

Alle in diesem Buch genannten Marken und Produktnamen unterliegen warenzeichen-, marken- oder patentrechtlichem Schutz bzw. sind Warenzeichen oder eingetragene Warenzeichen der jeweiligen Inhaber. Die Wiedergabe von Marken, Produktnamen, Gebrauchsnamen, Handelsnamen, Warenbezeichnungen u.s.w. in diesem Werk berechtigt auch ohne besondere Kennzeichnung nicht zu der Annahme, dass solche Namen im Sinne der Warenzeichen- und Markenschutzgesetzgebung als frei zu betrachten wären und daher von jedermann benutzt werden dürften.

Bibliographic information published by the Deutsche Nationalbibliothek: The Deutsche Nationalbibliothek lists this publication in the Deutsche Nationalbibliografie; detailed bibliographic data are available in the Internet at http://dnb.d-nb.de.

Any brand names and product names mentioned in this book are subject to trademark, brand or patent protection and are trademarks or registered trademarks of their respective holders. The use of brand names, product names, common names, trade names, product descriptions etc. even without a particular marking in this works is in no way to be construed to mean that such names may be regarded as unrestricted in respect of trademark and brand protection legislation and could thus be used by anyone.

Coverbild / Cover image: www.ingimage.com

Verlag / Publisher:
LAP LAMBERT Academic Publishing
ist ein Imprint der / is a trademark of
OmniScriptum GmbH & Co. KG
Heinrich-Böcking-Str. 6-8, 66121 Saarbrücken, Deutschland / Germany
Email: info@lap-publishing.com

Herstellung: siehe letzte Seite /
Printed at: see last page
ISBN: 978-3-659-11163-1

In vivo and ex vivo responses of leukemic cells to cladribine or
fludarabine combined with cyclophosphamide

Jolanta D. Żołnierczyk[1a], Arleta Borowiak[1a], Jerzy Z. Błoński[2], Barbara
Cebula-Obrzut[3], Małgorzata Rogalińska[1], Aleksandra Kotkowska[2], Ewa
Wawrzyniak[2], Piotr Smolewski[3], Tadeusz Robak[2], Zofia M. Kiliańska[1a]

[1]Department of Cytobiochemistry, Faculty of Biology and Environmental
Protection, University of Łódź, Pomorska 141/143, PL 90-236 Łódź,
Poland
[2]Department of Hematology, [3]Department of Experimental Hematology,
Medical University of Łódź, Ciołkowskiego 2, PL 93-510 Łódź, Poland

Correspondence: Zofia M. Kiliańska, e-mail: zkilian@biol.uni.lodz.pl

a – These authors have equal contribution and should be considered as
first Author

Subtitle: Drug-induced apoptosis of CLL cells

Table of contents

Abstract:

Background: The heterogeneity of chronic lymphocytic leukemia (CLL) is thought to be due to differences in the expression of factors that regulate apoptosis and cell cycle, giving rise to diverse apoptotic disturbances and tumor properties. Therefore, the primary goal in CLL treatment is to overcome resistance to apoptosis and efficiently trigger this process in leukemic cells.

Methods: Mononuclear cells were obtained from the blood of CLL patients by Histopaque-1077 sedimentation. CLL cell samples from the blood of drug treated patients (cladribine or fludarabine with cyclophosphamide; CC or FC), as well as the cell samples of untreated patients exposed to the used drug combinations (CM, FM) or mafosfamide alone for 48h were fractionated into nuclear and cytoplasmic fractions or were lysed. DNA fragmentation was evaluated by agarose electrophoresis and also cytometrically as sub-G1 population. The expression of apoptosis related proteins and H1.2 histone translocation were evaluated in lysates and nuclear and cytoplasmic fractions, respectively with appropriate antibodies.

Results: Cladribine (C) and fludarabine (F) combined with cyclophosphamide/mafosfamide *in vivo*, as well as *ex vivo* trigger apoptosis in CLL cells. These drug combinations (CC, FC, CM and FM) induce leukemic cell apoptosis confirmed by DNA fragmentation, sub-G1 cell number, down-regulation of anti-apoptotic proteins (Mcl-1, Bcl-2), and H1.2 histone translocation in comparison with appropriate control cells, however, to a different degree.

Conclusions: The kinetics and rate of drug-induced apoptosis in leukemic cells under *ex vivo* experiments differ between patients, mirroring the differences noticed during *in vivo* treatment. Individual model cell samples indicate comparable susceptibility to the used drug combinations under *in vivo* and *ex vivo* conditions.

Key words:
Chronic lymphocytic leukemia, cytotoxicity, cladribine, fludarabine, cyclophosphamide/mafosfamide, anti-leukemic therapy, apoptosis process, apoptosis-related proteins, histone H1.2 translocation, DNA fragmentation, sub –G1 cells, *ex vivo* and *in vivo* conditions

Abbreviations:
AIF – apoptosis inducinf factor, Apaf-1 – Apoptosis protease activating factor-1, ATM – ataxia telangiectasia mutated, Bak – Bcl-2-antagonist/killer, Bax – Bcl-2 associated X protein, Bcl-2 – B-cell leukemia/lymphoma-2 protein, C/2-CdA – cladribine/2-chloro-2'-deoxyadenosine, CC – cladribine + cyclophosphamide, CLL – chronic lymphocytic leukemia, CM – cladribine + mafosfamide, CR – complete remission, Ctr – control cells, EDTA – ethylene diamine tetraacetate, F – fludarabine monophosphate (the soluble form of fludarabine: 9-β-D-arabinosyl-2'-fluoroadenosine; F-ara-A), FC – fludarabine + cyclophosphamide, FISH – fluorescence *in situ* hybridization, FM – fludarabine + mafosfamide, IAP – inhibitory apoptosis protein, M – mafosfamide, Mcl-1 – myeloid cell leukemia-1, NR – non-responder, ORR – overall response rate, PARP-1 – poly(ADP-ribose) polymerase-1, PBS – phosphate buffered saline, PR – partial remission, SDS-PAGE – sodium dodecyl sulfate -- polyacrylamide gel electrophoresis, Smac/DIABLO – second mitochondrial activator of caspase/direct IAP binding protein with low pI

1. Introduction

Chronic lymphocytic leukemia (CLL) is the most frequent B-celll malignancy of adults in the western hemisphere which manifests a remarkably diverse clinical course [32]. The disease is diagnosed in individuals in their sixth decade of life with an incidence of 4.2/100 000/year. The incidence grows up to 30-40/100 000/year at an age of over 80 years [1, 11] . Some patients rapidly progressed and died within a few months after diagnosis whereas others have a long life. Approximetely one-third of patients survive over 20 years and never require therapy. It is suggested that 5-10% of CLL cases can develop an aggressive lymphoma during progression of the disease which originally is described as Richter syndrome [24]. Despite of years of research effort the cause of CLL remains unknown. The risk factors for the development of this hematological cancer have been identified, including advanced age, male sex, white ethnicity and inherited predispositions (5-10% or other hematological disorders [17]. A major feature of CLL patients is their susceptibility to reccurent infections, which are main cause of morbidity and mortality in this cancer. In majority of patients, the therapy should aim at obtaining complete remission as well as prolonging disease-free survival and overall survival. New therapeutic approaches, mainly immunochemotherapy, enable the improvement of the above parameters. However, this leukemia is still an incurable disease [37].

In the course of CLL an accumulation of immunologically incompetent B lymphocytes in the blood, bone marrow, and lymphoid organs takes place. This accumulation of leukemia cells may be caused by some disturbances in their apoptotic machinery [46, 47], the deregulation of cross-talk between cells and microenvironment [40, 59], as well as by an impairment in the cell proliferation [40,58]. The defects in the cell death pathways contribute not only to disease pathogenesis, but also to the development of resistance to cytotoxic drugs [23, 44]. Apoptosis in CLL cells is mainly driven through mitochondrial pathway [38]. In this pathway, the programmed cell death is induced by an intrinsic stimuli/signals which arrive at mitochondria causing loss of mitochondrial transmembrane potential and the release of the pro-apoptotic factors such as cytochrome c and apoptosis-inducing factor (AIF) into the cytosol. In the cytoplasm cytochrome c binds to Apaf-1 (Apoptosis protease activating factor-1) triggering an assembly of a large caspase-9 activating platform called as apoptosome [21]. Active caspase-9 initiates a proteolytic cascade responsible for the hydrolytic cleveage of important cellular proteins. According to the current knowledge, AIF is main mediator of caspase-independent cell death. This factor translocates from the cytosol to the nucleus and provokes effects of marginal chromatin condensation and large scale (50 kb) DNA fragmentation [26]. The common action of caspases, apoptosis-related exonucleases and AIF leads to cell shrinkage, blebbing, nuclear condensation and fragmentation, the formation of apoptotic bodies, and finally to their phagocytosis by macrophages or adjacent cells [2, 4, 12, s 28, 52, 55]. Fas receptor expression, and apoptosis induction *via* an extrinsic pathway practically does not occur in CLL cells [63].

Purine analogs – cladribine (C; 2-CdA) or fludarabine (F; the soluble monophosphate of F-ara-A) are used in monotherapy, as well as in conjunction with alkylating agent – cyclophosphamide in the current management of CLL [27,43, 48, 51]. Fludarabine (5' monophosphate of 9-beta-D-arabinosyl-2-fluoroadenine) is quantitatively dephosphorylated to F-ara-A immediately upon intravenous infusion. Cladribine (2-chloro-2'-deoxyadenosine) is a compound that resembles fludarabine. Process

10

of adenine ring 2-halogenation in both analogs induces the changes that make the amino group resistant to adenosine deaminase,, and results in their cytoplasmic accumulation under phosphorylated form, inducing cell death[15, 38, 62]. Preferential accumulation is observed in lymphoid cells since they contain a particularly high content of pyrimidine salvage pathway enzyme – deoxycytidine kinase [14]. The main differences between both above purine analogs are the presence in cladribine a chlorine atom on 2-carbon of the purine ring (fludarabine contains fluorine atom) and deoxyribose (fludarabine has arabinose). These diversities lead to the different dosing schedules, i.e. 20-30 mg/m^2 for fludarabine versus 4-5 mg/m^2 for cladribine. Purine analogs have been indicated to have a main effect on DNA synthesis through incorporation into DNA and interference with DNA polymerase and DNA ligase, and through inhibition of DNA primase, as well as on DNA repair. Both, fludarabine and cladribine are able to inhibit RNA synthesis, however , by the different manner. They are also potent to affect DNA methylation [14, 62].

Cyclophophosphamide – 2-[bis(2-chloroethyl)amino]tetrahydro-2H-1,3,2-oxazaphorine2-oxide monohydrate belongs to alkylating agents that indicates activity against a large number of cancers, including CLL [14]. The cytotoxic lesion is thought to result from inter-strand DNA cross-links, which are formed *in vitro* in the cells incubated with the activated prodrug - mafosfamide, or after *in vivo* cyclophosphamide application. Cyclophosphamide requires CYP450-dependent activation, yielding the active agent – 4hydroxycyclophosphamide. This compound undergoes immediately decomposition to active alkylating agent – phosphoramide mustard and acrolein. The former species alkylates the N7 of guanine. Both molecules, cyclophosphamide and *in vitro* mafosfamide reveal high potency in the triggering apoptosis of a large number of cell lines and solid tumors [18,39]. Cyclophosphamide/ mafosfamide synergizes with fludarabine or cladribine inducing cytotoxicity and apoptosis of lymphoid cells [3, 14].

The combining therapy for CLL based on cyclophosphamide with above purine analogs gives better efficacy suggesting that the formation of DNA cross-links after alkylating compound is prolonged with concomitant administration of fludarabine or cladribine, presumably these drugs inhibited DNA biosynthesis and repair. Both these purine analogs and the alkylator have been reported as potent agents when used in apoptosis induction and cell cycle inhibition [3, 38, 54, 66].

Little is known about the comparison of biological and clinicall activity *in vivo* and *ex vivo* of cladribine or fludarabine in combination with cyclophosphamide /mafosfamide.

In the current study, we aimed to evaluate apoptotic events and clinical activity in primary tumor cells isolated from blood of CLL patients treated with cladribine + cyclophosphamide or fludarabine + cyclophosphamide. The blood samples were isolated from the patients at four points during the course therapy; firstly, before the therapies were administered, then, after the first and third day of drug administration, and finally, two weeks after the end of the first cycle of treatment. Additionally, these values for *ex vivo* cytotoxicity and pro-apoptotic potential of the examined drugs were compared with those cells obtained from blood samples of patients before therapy.

2. Materials and Methods

2.1 Patients and treatment modality

The patients with previously untreated progressive CLL were enrolled onto the study. The diagnosis of CLL was established according to National Cancer Institute Sponsored Working Group (NCI-WG) criteria [9, 45]. CLL was identified as progressive in patients with Rai stage III and IV or patients with stage 0 - II [44] with any of the following symptoms: the progressive lymphocytosis (doubling time < 6 months), bulky lymphadenopathy or massive splenomegaly, weight loss above 10% over 6 months, disease-related fever $\geq 38^0$ C, or extreme fatigue.

For *in vivo* research, mononuclear cell samples obtained from fresh peripheral blood samples of 54 CLL patients treated with CC (cladribine + cyclophosphamide) or FC (fludarabine monophosphate + cyclophosphamide) regiments were taken. The group included eligible patients (from central Poland) who underwent randomization procedure [49]. The study was approved by the Local Ethics Committee of the Medical University of Łódź (No RNN/237/03KE) and all patients enrolled in the trials signed a consent form. All the patients showed the expression of CD5, CD19 and CD23 antigens on the surface of leukemic cells and monoclonality for the light chains of their immunoglobulins.

Protocol of CC treatment (32 patients) includes cladribine administered intravenously at dose of 0.12 mg/kg/day over 2 h daily for 3 days and cyclophosphamide administered at a dose of 250 mg/m^2 as continuous infusion for 3 days [48, 50].

The protocol of FC treatment (22 patients) includes fludarabine monophosphate administered intravenously at a dose of 25 mg/m^2 over 2 h daily for 3 days and cyclophosphamide administered at a dose of 250 mg/m^2 as continuous infusion for 3 days [51].

Cladribine (Biodrybin) was obtained from the Institute of Biotechnology and Antibiotics Bioton (Warsaw, Poland). Fludarabine monophosphate and cyclophosphamide were purchased from Schering AG (Berlin, Germany) and Asta Medica (Frankfurt, Germany), respectively.

2.2 Response criteria

The efficacy of treatment was assessed according to NCI-WG criteria [9]. A complete response (CR) was defined as the disappearance of tumor masses and disease-related symptoms and organomegaly, as well as normalization of initially abnormal parameters (absolute lymphocyte count < 4 × 10^9/l; neutrophil count >1.5 × 10^9/l; hemoglobin concentration > 110 g/l, and platelet count > 100 × 10^9/l; bone marrow that contained < 30% of lymphocytes for at least 2 months). Partial response (PR) was defined as a decrease ≥ 50% of measurable lesions. Patients not fulfilling the above criteria were considered as non-responders (NR).

2.3 Cell immunophenotyping

Immunotyping of CLL cells was determined in the whole peripherall blood samples. The routine panel of monoclonal antibodies (CD3,CD5, CD10, CD11c, CD19, CD23, FMC7, Ig kappa, Ig lambda), fluorescein isothiocyanate (FITC) , R-Phycoerythrin (R-PE) or cyanine-5 conjugated (all reagents from BD Pharmingen, San Diego, CA, USA) was performed using flow cytometry. The detecting the leukemic CD5+/ CD19+, CD23+ clone confirmed the diagnosis in all cases. The rate of CLL cell population ranged from 55% to 97% of perripheral blood white cells (mean 80.5%).

2.4 Isolation of CLL cells

The mononuclear cells of peripheral blood (PBMCs) were isolated in the Histopaque-1077 density gradient (Sigma-Aldrich, St. Louis, MO, USA). Peripheral blood samples (collected to EDTA) were obtained 1 day before the introduction of the treatment (day 0), during drug administration, i.e., after 1 and 3 days of CC and FC regiment. Additionally, blood samples were taken from patients 2 weeks after the last dose of drug administration as described previously [30, 66]. The CLL cell pellets obtained from blood of untreated or treated patients were then resuspended in phosphate buffered saline (PBS) and divided into the parts needed for the planned experiments.

2.5 Fluorescence *in situ* hybridization (FISH) analysis

FISH analysis was performed on interphase nuclei of leukemic cells on blood smears collected from patients before the introduction of the treatment as described previously [50]. Commercial probes (Vysis, Bergish-Gladbach, Germany) were used including the microsatellite chromosome 12 probe D12Z3 and the unique sequence – or region-specific DNA probes p53 (17p13.1.1. locus), ATM (11q 22.3 locus), and D13S319 (13q14.3 locus). Signals were counted in at least 200 interphase nuclei for each sample. To establish the cut-off level for the deletions and trisomy 12, experiments were made on peripheral blood smears from 10 healthy donors. The deletion was considered to occur if the specimen under study was greater than a mean value ± standard deviation of 3 nuclei with only 1 signal (referring to deletions) or with 3 signals (referring to trisomy 12).

2.6 Cell fractionation

The cell pellets were rinsed with cold PBS and then suspended in isotonic sucrose containing 5 mM $MgCl_2$, 0.5% Triton X-100, 50 mM Tris-HCl (pH 7.4) and protease inhibitors as described previously [53]. The cells were homogenized in a Potter homogenizer for 3 min at 80 V and filtered through several layers of gauze. A part of homogenate was left and kept at −20°C for further analysis and the rest was spun down at 800 × g for 7 min resulting in the crude nuclear pellet. The nuclear pellet was purified by sucrose method [5]. The post-nuclear fraction supernatant was designated as a crude cytoplasmic fraction.

2.7 Cell culture and *ex vivo* drug sensitivity testing

CLL cell samples isolated from blood of patients prior to the introduction of the therapy were taken for *ex vivo* research. Freshly obtained leukemic cells were left untreated (in RPMI medium; control cells; Ctr) or were exposed to: cladribine or fludarabine monophosphate or mafosfamide alone, or to combinations of cladribine with mafosfamide (CM) and fludarabine monophosphate with mafosfamide (FM) for 24 and 48 h. The concentrations of cladribine (C; 0.175 µM), fludarabine monophosphate (F; 20 µM) and mafosfamide (M; 2.5 µM) were used as described previously [2, 3, 7, 31]. All cell samples were maintained for 24 and 48 h in RPMI 1640 medium containing 10% (v/v) heat-inactivated fetal calf serum and antibiotics (streptomycin 100 μg/ml, penicillin 100 U/ml) at 37°C, 5% CO_2, fully humidified atmosphere. Mafosfamide was kindly donated by Baxter Oncology GmbH (Frankfurt, Germany).

The numbers of living cells were determined by flow cytometry (BD LSR II) using Vybrant Apoptosis Assay kit #4 according to manufacturer's instruction (Invitrogen Molecular Probes Inc., Eugene, USA) as was previously described [55]. Briefly, the cell samples were simultaneously stained with green fluorescent Yo-Pro-1 and red fluorescent propidium iodide (PI) dyes. Cellular accumulation of the fluorochromes was measured at two channels and evaluated by a bivariate analysis. The distinction between living (YO-PRO$^-$/PI$^-$), apoptotic (YO-PRO$^+$/PI$^-$) and dead (YO-PRO$^+$/PI$^+$) cells was based on differences in the permeability of their cellular membranes for fluorescent dyes. The results of analysis by Vybrant Apoptosis Assay kit #4 are shown as the percent of viable cells in the mononuclear cell population. The measurement was carried out before the exposure as well as after 24-h and 48-h exposure of cell samples to the tested agents.

2.8 DNA fragmentation

Apoptotic cell fragmentation was evaluated on the basis of DNA content in cell nuclei in the population of CLL cells isolated from blood of patients [19,22, 26, 42, 61]. Ethanol-fixed mononuclear cells were stained with propidium iodide (450 μl PI solution at concentration of 50 μg/ml per 1×10^6 cells) in the presence of RNase (50 μl RNase solution at concentration of 100 μg/ml per 1×10^6 cells) and analyzed by flow cytometer (FACS Calibur BD, San Jose, CA, USA) equipped with 488 nm light source. Leukemic cells with oligonucleosomal DNA fragmentation were visualized on DNA content histograms as a hypodiploid DNA peak ("sub-G1 population"). Number of sub-G1 cells was evaluated using Cell Quest Pro Software (Becton Dickinson). Ten thousand events were examined for each analysis.

DNA fragmentation was also evaluated by agarose electophoresis performed according to a slightly modified procedure described previously [30]. Briefly, after washing with PBS, 6 × 10^6 CLL cells were lysed and treated with proteinase K (0.2 mg/ml) in a buffer containing 5 mM Tris-HCl, pH 8.0, 20 mM EDTA, 0.5% Triton X-100 for about 12 h at 37°C. DNA was extracted twice with buffered phenol/chloroform/isoamyl alcohol (25 : 24 : 1, v/v/v), and precipitated with 0.1 volume of 3 M sodium acetate and 2 volumes of ethanol at −20°C, overnight. DNA precipitates were washed twice with 75% ethanol, dissolved in triple-distilled water, and digested with RNase A (1 mg/ml) for 2 h at 37°C. Finally, DNA samples were electrophoresed by standard agarose gel (2.0%) electrophoresis. Puc 18 DNA Hae III digest samples (used as marker) were run in the same gels. The gels were stained with 10 µg/ml ethidium bromide. To visualize apoptotic alterations in DNA integrity, the gels were observed under UV.

2.9 Preparation of whole cell lysates

Control and drug-treated CLL cells were lysed (4°C; 20 min) in a buffer containing 10 mM Tris-HCl (pH 7.5), 300 mM NaCl, 1% Triton X-100, 2 mM $MgCl_2$, 0.1 M dithiothreitol, and protease inhibitors as described previously [66]. Protein content was estimated by method of Lowry et al. [36].

2.10 SDS-polyacrylamide gel electrophoresis and immunoblotting assay

Protein concentration in the homogenates, cellular fractions and whole lysates of CLL cells were determined colorimetrically [36] and prepared for subsequent Western blotting analysis as previously described [30]. The samples (50 µg/lane) were separated by SDS-polyacrylamide gel electrophoresis (SDS-PAGE) on 8.0 or 12.5% slab gels and electrotransfered onto Immobilon P [33, 60]. Equal protein loading and completeness of the transfer were checked by membrane staining with 0.5% Ponceau S solution. The membranes were saturated in 5.0% skim milk in TBS (10 mM Tris-HCl, pH 7.5, 150 mM NaCl) for 1 h at ambient temperature and then incubated overnight with antibodies specific to: Mcl-1 (sc-819; 1 : 1000), Bcl-2 (sc-787; 1 : 1000), Bax (sc-493; 1 : 1000), caspase-9 (sc-8355; 1 : 4000), lamin B (sc-6216; :1000), PARP-1 (sc-7150; 1 : 1000) as well as Bak (Ab 2273; 1 : 1000) and histone H1.2 (Ab 17677; 1 : 1000) from Santa Cruz Biotechnology Inc. CA, USA and Abcam Ltd. Cambridge, UK, respectively. In some experiments antibodies to Mcl-1 (catalog # 444206) and caspase-9 (catalog # 218794) were from Oncogene Research Products, CA, USA. Subsequently, antigen recognition was performed with proper secondary antiserum conjugated with alkaline phosphatase. The immune-complexes were detected after incubation with the substrates as previously described [31, 34]. Moreover, antibody to actin (Ab 1801; 1:1000) (Abcam Ltd. Cambridge, USA) was used as additional loading control.

2.11 Statistical analysis

The experimental data were expressed as the means ± standard deviations. U Mann- Whitney test and Wilcoxon signed rank test were carried out to determine significant differences. Probability values $p < 0.05$ were considered to be statistically significant.

3. Results

The blood samples of 54 patients with diagnosed CLL [39 men (72.2%) and 15 women (27.8%)] were used in *in vivo* study. Table 1 summarizes the clinical characteristics of the patients. All the patients had undergone a course of therapy (after randomization procedure) based on a combination of cladribine or fludarabine monophosphate with cyclophosphamide, i.e., CC (22 men and 10 women) or FC (17 men and 5 women) [48, 51].

Tab. 1. Characteristic and outcome of treatment with CC and FC in individual patients

Protocol CC						
No.	Patient initials	Age	Leukocytosis × 10^9/l	Stage of disease	FISH results	Response to treatment
Male						
1	TH	81	62	I	12	PR
2	KM	51	140	II	N	CR
3	KD	59	59	II	–	PR
4	WR	63	27	I	11	CR
5	KK	73	190	II	N	STOP

20

No	Patient initials	Age	Leukocytosis x10^9/L	Stage of disease	FISH results	Response to treatment
6	WS	51	44	I	11	PR
7	KK	76	191	I	13	CR
8	GA˙	53	330	IV	11	NR
9	CA	60	100	IV	N	PR
10	NG	60	46	IV	12	PR
11	KM˙	72	111	I	12	CR
12	KS˙	64	117	I	11	CR
13	KH˙	77	117	II	12	PR
14	MJ˙	67	180	0	13	CR
15	BK˙	76	160	IV	11, 13	NR
16	KR˙	56	160	IV	13	NR
17	ŁJ˙	62	256	II	11	CR
18	SA˙	55	138	IV	N	CR
19	GA˙	52	110	II	11, 13	CR
20	KM˙	69	207	II	13	CR
21	BM˙	51	120	IV	11, 12	CR
22	ŚJ˙	69	300	IV	–	CR
Female						
1	NL	70	300	IV	11, 12	PR
2	PK	55	27	I	–	PR
3	PK˙	72	122	0	–	PR
4	PK	76	69	III	11	CR
5	PA˙	75	130	III	11, 17	PR
6	JZ˙	79	135	III	N	CR
7	RB˙	50	180	IV	11	CR
8	SI˙	52	105	I	11, 12, 13	PR
9	SG	49	319	II	–	NR
10	JB˙	55	129	II	–	CR

Protocol FC

No	Patient initials	Age	Leukocytosis x10^9/L	Stage of disease	FISH results	Response to treatment
Male						
1	WS	47	37	IV	–	STOP
2	SS	72	102	IV	12	NR
3	ŚW	52	91	IV	N	PR
4	WM	46	176	I	–	PR
5	PK	47	45	III	–	STOP
6	ŁS˙	60	142	II	17	CR

7	OJ*	65	100	IV	N	STOP
8	KS	55	80	III	N	NR
9	RK*	56	82	IV	–	CR
10	OE*	76	100	IV	–	CR
11	BM*	66	80	I	12	CR
12	KM*	61	200	II	11	PR
13	LH*	62	70	I	11, 13	CR
14	CJ	58	68	I	N	CR
15	ZJ*	65	91	III	11	STOP
16	UZ*	63	300	IV	11	STOP
17	SS*	71	180	IV	11	CR
Female						
1	RR	71	106	II	12	CR
2	JE*	40	88	II	12, 13	CR
3	DG*	78	140	I	13	CR
4	RG*	75	75	IV	N	CR
5	DT*	65	100	0	11	PR

The clinical staging of CLL was determined according to the Rai et al. [32]; CR – complete response, PR – partial response, NR – nonresponder, – = not determined

Most of the patients received 6 cycles of chemotherapy; in 6 cases the treatment had to be stopped. Additionally, we carried out the experiments *ex vivo* in which CLL cells were obtained from the same patients before the therapy (Tab. 1, bolded cases with asterisks), when such materials were available. CLL cell samples isolated from blood of 33 patients before drug administration (day 0) were exposed to cladribine, fludarabine monophosphate, mafosfamide (active form of cyclophosphamide *ex vivo*) alone as well as to CM and FM combinations at pharmacological concentration for 24 and 48 h as was described previously [65]. Simultaneously, CLL cell samples were incubated in the culture medium without any drugs (control).

3.1 Cell viability and apoptosis induction *ex vivo*

For *ex vivo* analyses, the blood samples were obtained from CLL patients before the administration of anticancer agent(s). The viability and the numbers of apoptotic cells in leukemic cells isolated from blood and exposed to tested drugs *ex vivo* were studied cytometrically by Vybrant Apoptosis Assay #4. The results of drug cytotoxicity for patients (15 and 17, who received CC or FC on the next day, respectively) are presented in Figure 1.

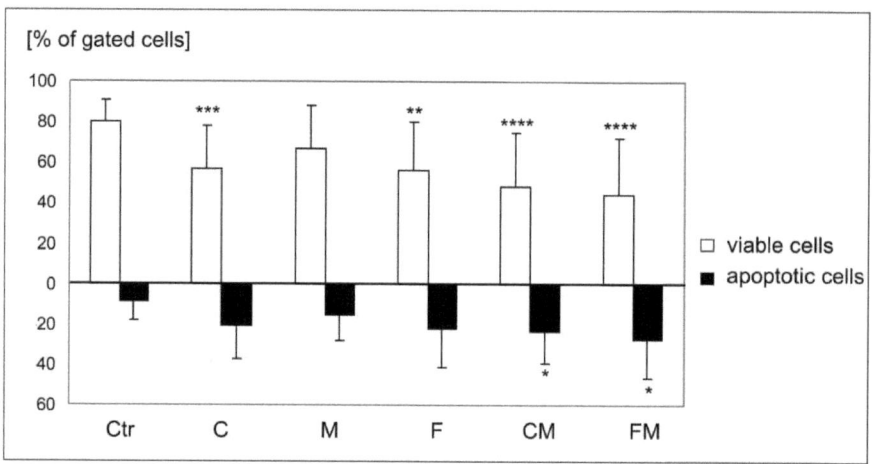

Fig. 1. Viability and apoptosis rate of CLL cells isolated from the blood of patients before therapy exposed to cladribine (C), mafosfamide (M), fludarabine monophosphate (F) and combinations of above purine analogs with mafosfamide – CM and FM as well as the cells incubated in culture medium without any agents (control, Ctr) for 48 h. The data represent the mean values ± SD from 13–16 experiments of CLL cell

exposure to C, M, F, and 12–14 to CM and FM combinations, respectively. Asterisks indicate statistical significance: * $p < 0.002$, ** $p < 0.001$, *** $p < 0.0005$, and **** $p < 0.0001$

A 24-hour exposure of model cells to the tested drugs used solely did not change essentially the viability of the model cells (data not shown). After 48 h, both, cladribine and fludarabine monophosphate, as well as their combinations with mafosfamide, significantly influenced cell viability and the number of apoptotic cells in the tested population of CLL cells. However, the differences in the cytotoxic and apoptotic potential of drugs individually introduced to culture media and their combinations were clearly observed. The combined action of purine analogs and mafosfamide caused a decrease of viable cell number at a statistically significant level ($p < 0.0001$; $U = 202$ and $p < 0.0001$; $U = 201$ for CM and FM, respectively; U Mann-Whitney test) in comparison to the control. However, the differences in the cytotoxic and apoptotic potential of drugs individually introduced to culture media and their combinations were clearly observed. FM and CM provoked the highest decline of cell viability, which was also accompanied by the growth of apoptotic cell numbers (Fig. 1). The percentage of dead cells (YO-PRO+/PI+) rose proportionally to the degree of apoptosis induction (data not shown).

3.2 Sub-G1 cell population and DNA ladder formation *in vivo* and *ex vivo*

To investigate the pro-apoptotic and clinical activity of the CC and FC regimens on CLL cells *in vivo,* the populations of primary tumor cells isolated from blood of 32 CC-treated and 22 FC-treated CLL patients were examined. The blood samples were obtained prior to the introduction of the therapy (0), after day 1 and 3 of treatment, as well as two weeks after the first cycle of drugs administration (day 17). The methodology used was based on cytometric assessment of DNA content histograms, and electrophoretical analysis of DNA fragmentation in the tested leukemic cells.

The numbers of hypodiploid events, later called "sub-G1 cell population" were studied, among all the events noticed on DNA content histograms derived from lymphocyte-gated plots. It is believed that "sub-G1" population corresponds with the population of apoptotic cells characterized by oligonucleosomal DNA fragmentation.

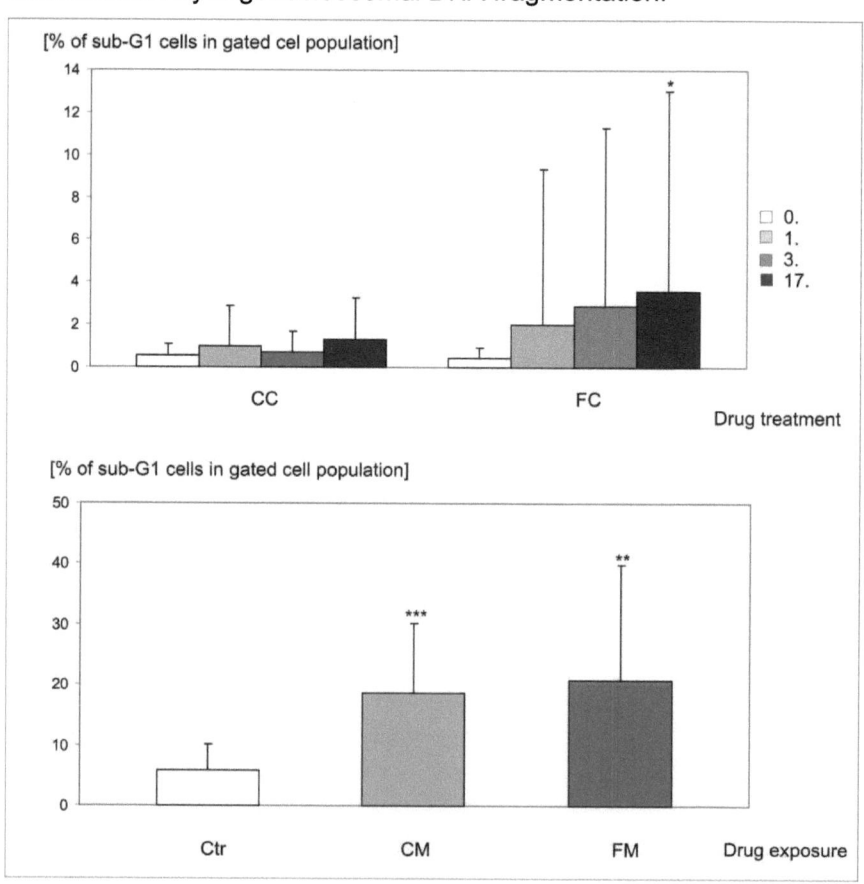

Fig. 2. Level of sub-G1 cells in CLL samples after *in vivo* and *ex vivo* treatment with cladribine or fludarabine monophosphate combined with cyclophosphamide/mafosfamide. Numbers of sub-G1 cells in CLL cell samples obtained from blood of patients before therapy (0), and after the first (1), and third day (3), as well as 14 days (17) after the initial chemotherapy cycle with CC and FC (**A**), and in the cell samples isolated from blood of patients before drug treatment and exposed to CM and FM for 48 h (**B**). Control cells (Ctr) were incubated in RPMI culture medium without any drugs. The data represent the mean values ± SD from 15 (CC) and 17 (FC), experiments carried out *in vivo* and 15 (CM) and 14 (FM) experiments *ex vivo*, respectively. Asterisks indicate statistical significance: * $p < 0.05$, ** $p < 0.002$, and *** $p < 0.0001$

Our results (Fig. 2A) revealed that both CC and FC therapeutic options induce small increase of sub-G_1 cell count in drug-treated cells versus control – CLL cells isolated from blood of the patients prior to the therapy (statistic significance was only noted   after FC therapy; the Wilcoxon signed-rank test, $p < 0.05$). As shown in Figure 2A, the highest effect was observed two weeks after the last dose of the initial cycle of the CC and FC therapy (day 17), when the number of sub-G1 cells increased to over 2.4 and 8.7 times, respectively, in relation to the control.

To confirm whether DNA fragmentation caused by the examined drug combinations resulted from cleavage between nucleosomes, DNA samples were separated by agarose gel electrophoresis. As it is shown in Figure 3A, agarose gel electrophoregrams of DNA obtained from equal number of examined cells indicate the increasing amounts of oligonucleosomal DNA fragments (apoptotic ladder) after the drug treatment. Interestingly, the formation of the ladder was noticed after day 1 of drug administration, and was enhanced during the application of drugs.

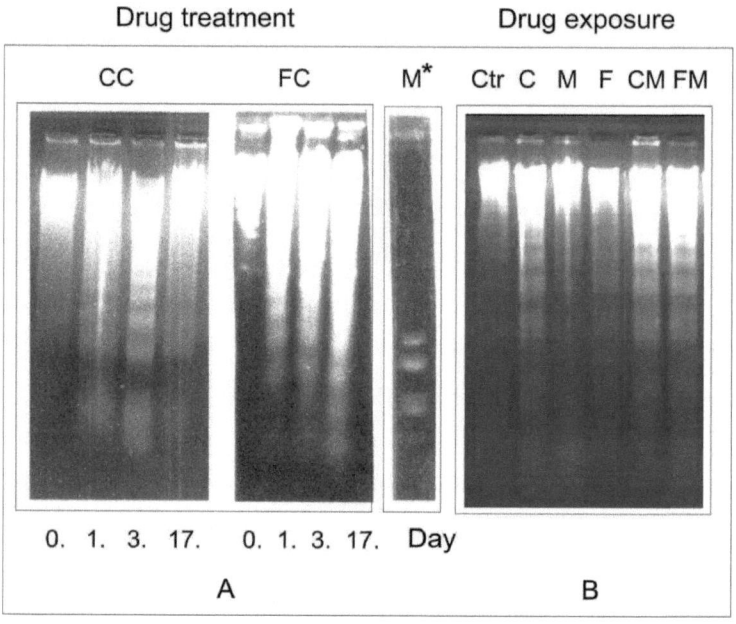

Fig. 3. DNA ladder formation in CLL cell samples after *in vivo* (A) and *ex vivo* (B) treatment with cladribine or fludarabine/fludarabine monophosphate combined with cyclophosphamide/mafosfamide as well as with the drugs used alone. DNA was isolated from CLL cells of exemplary patients before therapy (0) and administered with CC (KM, No. 11) and FC (LH, No. 13) and from the cell sample (patient No. 153 before therapy) after exposure to cladribine (C), mafosfamide (M), fludarabine (F), CM and FM combinations (B). Lane M* DNA markers; Ctr, DNA from PBMC incubated in RPMI culture medium without any drugs

The results of DNA content analysis, illustrated in Figure 2B indicated elevated numbers of sub-G1 cells in leukemic cells exposed to drug combinations, in comparison with the control cells. Interestingly, extensive oligonucleosomal DNA fragmentation was noticed in the electrophoregrams of DNA mainly from CLL cells that had been incubated with CM and FM (Fig. 3B).

3.3 Expression of apoptosis-related proteins and histone H1.2

To further examine the other apoptotic events, the expression of chosen apoptosis-related proteins in CLL cells was determined by western blot technique. The samples of the nuclear and cytoplasmic fractions of the blood of patients who were treated according to CC or FC protocol were examined. The expression level was evaluated before therapy (day 0), after day 1 and 3 of the first cycle of chemotherapy and 14 days (day 17) after the final drug application. Several apoptosis-related proteins were analyzed in almost all studied cases. The immunoblotting results of two exemplary patients, who received the CC and the FC and responded to therapy (Tab. 1, patients No. 11, male, KM and No. 13, male LH) are presented in Figure 4. Both, CC and FC treatment caused clear changes in the expression level of two anti-apoptotic regulatory Bcl-2 family members – Mcl-1 and Bcl-2.

A significant drop in Mcl-1 expression was noticed in the cells of drug administered patients after day 1. On the contrary, the expression level of Bcl-2 did not significantly change after day 1 and 3, but a decreased level of this inhibitory protein was usually observed 2 weeks after the initial cycle of therapy. Interestingly, simultaneously with the decrease of the anti-apoptotic polypeptide level, an increase in the expression of Bax and Bak, the pro-apoptotic members of Bcl-2 family, was seen (Fig. 4).

The panel of apoptotic machinery proteins, including procaspase-9 and the nuclear proteins PARP-1 and lamin B was extended.

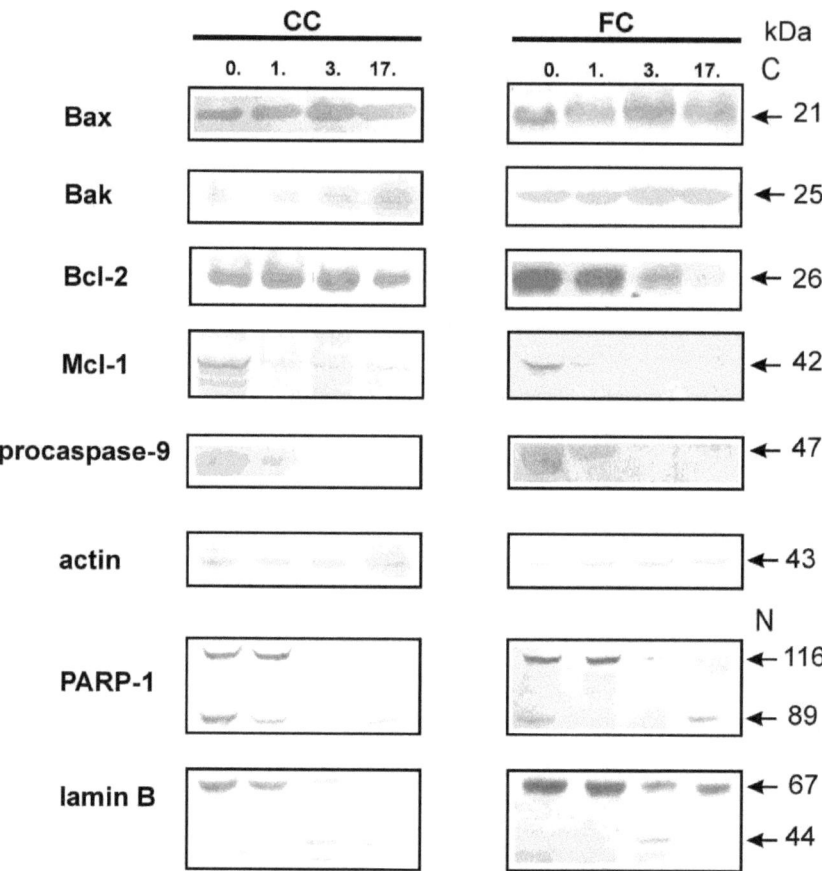

Fig. 4. *In vivo* changes in expression level of selected apoptosis-related proteins. Protein samples (40 µg) of cytoplasmic (C) and nuclear (N) fractions isolated from CLL cells of exemplary patients (KM, No. 11; protocol CC) and (LH, No. 13, protocol FC) were separated on 8.0 or 12.5% SDS-PAGE gels, transferred to a Immobilon P membrane, and analyzed for Bax, Bak, Bcl-2, Mcl-1, procaspase-9, PARP-1 and lamin B

by Western blotting. Actin expression was used as loading control. For details see Fig. 2A

It was revealed that the level of the caspase-9 precursor that was seen as an intensive band before the course of therapy, dropped quickly after day 3 of the drug administration. In the cells from the blood of the patients receiving CC and FC combinations, considerable proteolysis of chosen nuclear proteins occurred, although to a different degree. A deeper analysis of a large number of immunoblots revealed intra-patient differences in the expression of the investigated proteins in the samples after drug treatment.

The expression and proteolysis profile of the chosen apoptosis-related proteins were examined to identify molecular changes underlying the *ex vivo* apoptosis of drug-treated CLL cells. After the incubation (48 h) of leukemic cells with the tested agents and their combinations, cell lysates were assessed by immunoblotting. As is illustrated in Figure 5 (exemplary patient No. 21, male, BM, who achieved CR after CC treatment) Mcl-1, and in a lower extent, Bcl-2 protein, were down-regulated following treatment with both purine analogs, and their combinations with mafosfamide, in comparison with the control cells. The decrease of the Mcl-1 precursor was accompanied by an escalation of its 30 kDa C-terminal cleavage product. The expression level of the pro-apoptotic proteins Bax and Bak was usually elevated after drug treatment, especially after the exposure of cells to the drug combinations. Additionally, the expression level/cleavage of procaspase-9, an apical enzyme of the mitochondrial apoptosis pathway, was determined in the lysates of cells incubated in the presence of the investigated drugs. The immunoblot results demonstrated the occurrence of the proteolytic cleavage of procaspase-9 (47 kDa), accompanied by the appearance of 35/37 kDa product, which was intensively immunostained by probes incubated with fludarabine monophosphate, CM and FM compared with the untreated cells.

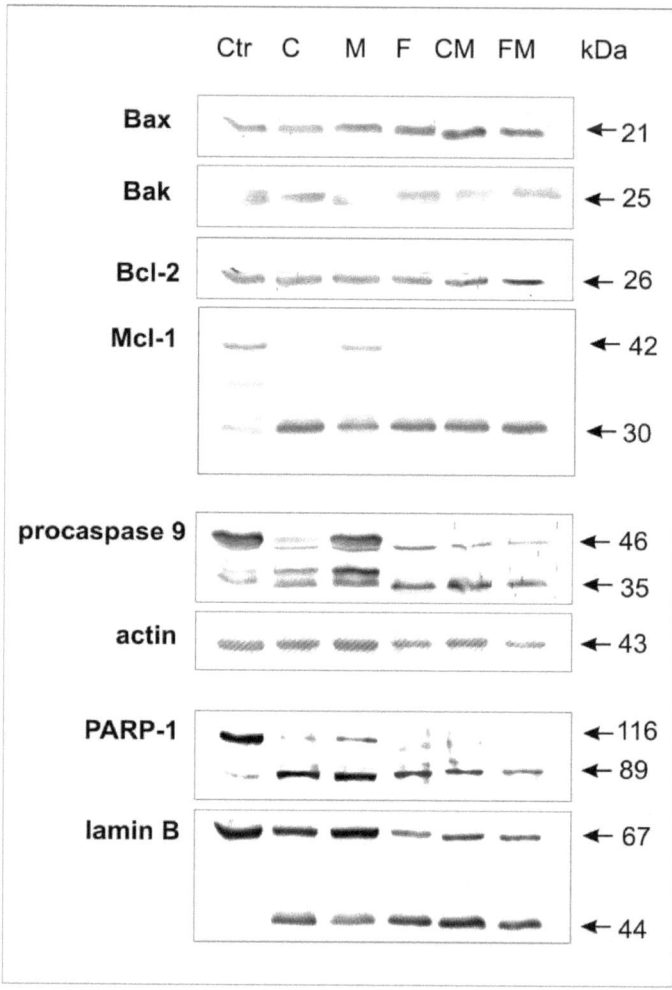

Fig. 5. *Ex vivo* changes in expression level of selected apoptosis-related proteins. CLL cell samples from blood of exemplary patient before

therapy (BM, No. 21), who responded to CC therapy, were exposed to C, M, F, CM, and FM for 48 h. Protein lysates (40 µg) from untreated (Ctr) and drug-treated CLL cells were separated on 8.0 or 12.5% SDS-PAGE gels. After electrophoretic separation and transfer, proteins immobilized on Immobilon P were analyzed for Bax, Bak, Mcl-1, procaspase-9, PARP-1 and lamin B by Western blotting. Actin expression was used as loading control

The data presented in Figure 5 revealed that in drug-treated cells, considerable proteolysis of PARP-1 and lamin B was also occurred. The precursor of PARP-1 (116 kDa) diminished or almost disappeared (CM and FM), and proteolytic product of 89 kDa emerged. A full length lamin B (67 kDa) was intensively proteolysed in drug-exposed cells, especially in the case of fludarabine monophosphate, CM and FM.

In this study, the translocation of histone H1.2 subtype between cellular fractions obtained from CLL cells was evaluated in the blood of the cured patients. This translocation has been recently described as a key event for apoptosis [16, 35]. It was confirmed that histone H1.2 was released from the cell nuclei to the cytoplasm in the cells of drug administered patients. The above histone H1 subtype was detected in leukemic cell nuclei (or homogenate, data not shown), whereas no (or slight) staining was noticed in cytoplasmic samples before therapy (Fig. 6A). Interestingly, in the cytoplasmic fraction of the studied cells from blood of patients who had undergone the therapy, the band stained with anti-histone H1.2 antibody had a different intensity (Fig. 6A).

The presence of histone H1.2 in the cytoplasmic fraction of leukemic cells exposed to apoptosis inductors was also observed. Figure 6B illustrates the presence of histone H1.2 in the nuclear and cytoplasmic fractions obtained from the blood of an exemplary patient

(No. 22, male, ŚJ) who was administered with CC and reached CR (Tab. 1). As is shown in Figure 6B, an intensive stained band recognized by anti-histone H1.2 antibody was seen in the cytoplasmic fraction of the examined cell samples treated with C, M and the combination of these components (CM). This histone H1 subtype was effectively released from the nucleus to the cytoplasm after CLL cell exposure to C or CM.

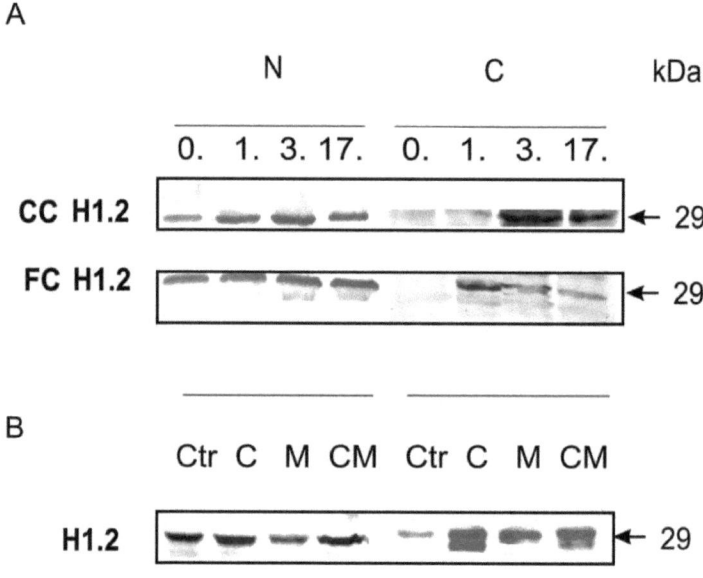

Fig. 6. Histone H1.2 translocation from nuclei to cytoplasmic fraction in CLL cells after *in vivo* (**A**) and *ex vivo* (**B**) treatment with cladribine or fludarabine monophosphate combined with cyclophosphamide/ mafosfamide. Forty micrograms of nuclear (N) or cytoplasmic fraction (C) isolated from CLL cells of exemplary patients administered with CC (KS, No. 12) or FC (ZJ, No. 15), respectively, and from the samples of patient before therapy (ŚJ, No. 22, protocol CC, who reached CR) exposed to cladribine (C), mafosfamide (M) and CM were separated on

12.5% SDS-PAGE gels, transferred to a Immobilon P membrane, and analyzed for histone H1.2 by western blotting. For details see Figure 2A; Ctr indicates control cells incubated in RPMI culture medium without drugs

3.4 The comparison of *in vivo* and *ex vivo* treatment results

An early link between the clinical responses of cured patients, *ex vivo* cytotoxicity and apoptotic activity was observed. Figure 7 shows representative results from two patients (No. 18, male, SA and No. 16, male, KR), who were both administered with CC (after randomization) but responded in an opposite manner (CR and NR). In the cell samples from the blood of a patient who reached CR after 6 cycles, a higher percentage of apoptotic cells and profound changes in the expression of selected apoptosis-related proteins were detected after 48 h cell incubation with drugs in comparison with the controls. The expression of pro-apoptotic regulatory proteins – Bax and Bak increased after CLL cell exposure to CM, comparing to control cells. Interestingly, Mcl-1 precursor almost disappeared, concomitantly, with the appearance of intensively stained proteolytic product with mol. wt. of 30 kDa. *Ex vivo* CLL cells of this patient were also sensitive to FM exposure (Fig. 7A). In contrast, *ex vivo* results confirmed that in CLL cells from blood of a patient, who did not respond to CC therapy, no influence on chemoresistance and the expression of the examined proteins was seen (Fig. 7B). The immunoblot analysis of control cells of these patients showed a different basal Bax expression, much higher in the cells of patient who did not respond to the therapy. The incubation of cell samples from blood of non-responder (patient No. 16) with CM did not change the expression of this pro-apoptotic protein. On the other hand,

CLL cells from blood of patient No. 18, who reached CR after therapy, were characterized by a higher Bax expression after their exposure to CM in comparison with control ones.

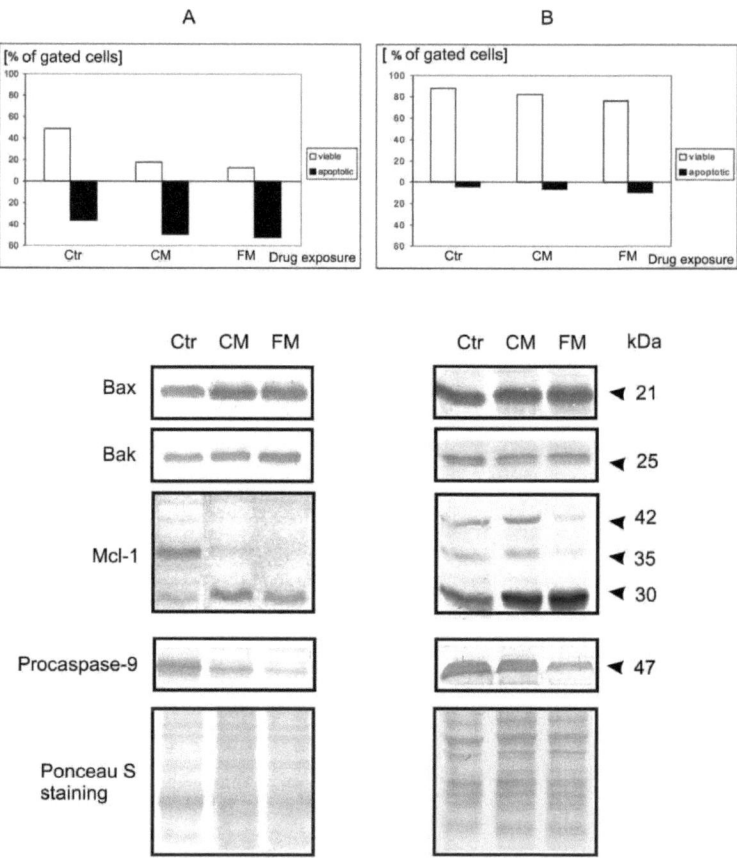

Fig. 7. Predictive value of *ex vivo* testing. The cell viability and expression level of selected apoptosis-related proteins of CLL cell samples isolated from exemplary patients who after randomization were administered with CC and achieved opposite response (SA, No. 18 CR;

A), (KR, No. 16, NR; **B**). The viability of studied cell samples exposed to CM and FM were shown after 48 h. Immunoblotting results indicate the expression of selected proteins: Bax, Bak, Mcl-1, and procaspase-9 in the lysates of CLL cells exposed to CM and FM for 48 h. Ponceau S staining was used as loading control; Ctr indicates control cells incubated in RPMI culture medium without drugs

It seems that *ex vivo* testing of leukemic cells with the chemotherapeutic(s) might be helpful in choosing optimal therapy options.

3.5 Treatment efficacy

The total numbers of randomized patients who were administered with CC or FC were 32 and 22, respectively. In all, 27 of 32 and 15 of 22 patients included in this study responded to CC and FC regimen, yielding a significant overall response rate (ORR) of 84.4 and 68.2%, respectively. Complete response (CR) was confirmed in over 50% of the cured patients, i.e., 16 who received CC and 11 who had received FC. In both groups, 6 patients showed no response to the applied therapy. The course of therapy had to be stopped in 6 patients, i.e., 1 in the CC group and in 5 in the FC group (Tab. 1). In the current study, the genetic abnormalities in most patients (i.e., 43) were determined by FISH technique. The occurrences of del 11q22, del 13q14, trisomy 12, and del 17p13.1 were detected, as well as normal karyotypes.

As shown in Table 2, del(11q) was the most frequent genetic abnormality followed by trisomy of chromosome 12 among patients who underwent therapy and reached ORR. It is interesting to note that among 43 tested patients, del(17p) was noticed only in 2 cases (4.65%).

Tab. 2. Chromosomal abnormalities and outcome of CLL patient treatment with CC and FC regimens, respectively

Response to treatment	n=26 Protocol CC			n=17 Protocol FC		
	Number of patients (%)		Chromosomal abnormalities	Number of patients (%)		Chromosomal abnormalities
Complete (CR)	14 (53.9)	5	Del 11q22	9 (52.9)	1	Del 11q22
		3	Del 13q14		1	Del 13q14
		1	Del 11q22, Del 13q14		1	Del 17p13.1
		1	Del 11q22, Trisomy 12		1	Del 11q22, Del 13q14
		1	Trisomy 12		1	Del 13q14, Trisomy 12
		3	Normal karyotype		2	Trisomy 12
					2	Normal karyotype
Partial (PR)	8 (30.8)	1	Del 11q22	3 (17.65)	2	Del 11q22
		1	Del 11q22, Del 17p13.1		1	Normal karyotype
		1	Del 11q22, Trisomy 12			
		1	Del 11q22, Del 13q14, Trisomy 12			
		3	Trisomy 12			
		1	Normal karyotype			
No response (NR)	3 (11.5)	1	Del 11q22	2 (11.8)	1	Trisomy 12
		1	Del 13q14		1	Normal karyotype
		1	Del 11q22, Del 13q14			
Stop	1 (3.8)	1	Normal karyotype	3 (17.65)	2	Del 11q22
					1	Normal karyotype

4. Discussion

Chronic lymphocytic leukemia is often considered to be a disease resulting from disturbed apoptosis [15, 38, 46, 57, 58, 64]. The resistance to apoptosis may stem from a combination of pro-survival signals from the microenvironment, as well as from intrinsic alteration in the apoptotic machinery of leukemic cells [8,47, 59, 47]. A subset of the primary tumor cells residing in the germinal centers is mitotically active [29]. Malignant cells escape apoptosis by a number of mechanisms, among which overexpression of anti-apoptotic genes has been reported to play a key role [13, 20, 23, 44]. It is now well established that signal transduction pathways involved in drug-induced apoptosis converge on a common pathway that consists of regulatory, effector and adapter molecules as Bcl-2 and IAP (inhibitory apoptosis protein) family, caspases, and Apaf-1 protein [6, 20,47]. The diversities between CLL patients in the expression and functional status of numerous anti-apoptotic and pro-apoptotic factors seem to be a main cause of the strong heterogeneity in disease development and in response to anti-leukemic treatment. More than 90% of CLL cells with typical hypercondensed heterochromatin are arrested in G0/G1 phase of cell cycle. According to teoretical predictions, the drugs with the capability to induce apoptosis process should be effective for irreversible elimination of leukemic cells by apoptosis [28, 46, 55]. Standard strategies currently used in clinical practice are based on combinating of purine analogs with alkylators (e.g. cyclophosphamide) that iduce DNA damage [14, 18]. They are applied in order to evoke cellular response to DNA injury. DNA breakage results in an activation of P53 protein, and as a consequence of apoptosis induction. This strategy, however, has an important limitation. It works only in the cells harbouring functional P53. It is accepted, that in numerous cancer sufferers *P53* is mutated. In CLL patients, the disruption of P53 pathway by *P53* mutation (about 10%) and by 17p13 deletion (3-7%) have been associated with a poor prognosis [65]. During CLL progression, some patients acquire resistance to purine derivatives [27, 44, 57]. Moreover, cladribine and fludarabine combined with cyclophosphamide seem to be sensitive for cycling cells, a minor part of CLL cells [38, 50].

In order to study the biological and clinical relevance of some *ex vivo* findings for the effects of anti-leukemic therapy *in vivo*, the response of CLL patients to treatment based on a combination of cladribine or fludarabine monophosphate with cyclophosphamide was evaluated. The obtained *in vivo* results revealed in most cases of CC and FC administration, a rapid decrease or even loss of Mcl-1 protein, usually after the first application of the drugs. After treatment, the expression level of Bcl-2, another anti-apoptotic protein, also declined, however, usually after the second dose of drugs. Both these Bcl-2 family members play a principal role in CLL cell apoptosis inhibition [6,10]. They are responsible for long lifespan of primary tumor cells and the resistance to the therapy [44, 57]. Their cellular function relies mainly on mitochondrial outer membrane permeability inhibition, due to neutralization of pro-apoptotic Bcl-2 family members [6, 10]. The falling expression of Mcl-1 and Bcl-2 was accompanied by the increase of Bak and Bax expression. Rapid diminution of Mcl-1 expression seems to procede proteolysis of procaspase-9 and nuclear proteins, i.e., PARP-1 and lamin B in model cells isolated from the blood of patients who were receiving CC or FC treatment. It is suggested that Mcl-1 is able to hamper apoptosis in hematopoietic cells to a greater extent than Bcl-2, and it displays a predominant role in the survival of transformed cells [25]. Its overexpression in CLL cells is thought to be associated with chemoresistance, as well as with the progression of the disease, whereas its decrease after exposure to treatment is a prerequisite for apoptosis induction. Additionally, it is thought that the appearance of 30 kDa cleavage product of Mcl-1 may act in positive feedback manner as a pro-apoptotic protein co-operating with Bax or Bak, both being promoters of apoptosis process [41].

Although it is hard to trace the course of apoptosis *in vivo*, both CC and FC regimens were found to trigger apoptosis in CLL patient cells. This was confirmed by the degree of oligonucleosomal DNA fragmentation, sub-G1 cell number, DNA ladder formation, down-regulation of anti-apoptotic proteins and H1.2 histone translocation seen during drug administration, in comparison with those of CLL cells from the blood of patients prior to therapy. Nevertheless, no significant global differences were observed in the intensity of the pro-apoptotic potential of both regimens. It appears, however, that cell samples isolated from blood of individual patients responded to therapy with high discrepancy, as evidenced by elevated standard deviation values in the tested groups.

Little information is available on the clinical responses to CC and FC in CLL patients in relation to genetic information. In contrast to many other countries, in Poland, the CC regimen is more frequently used than FC for treating this form of cancer [48,50]. In this study, it is revealed that a relative high proportion of CLL patients, previously untreated, respond to CC: the ORR reached over 84%, compared to over 70% for those receiving FC (Tab. 2). The obtained data reveal that a structural aberration of chromosome 11 followed by trisomy of chromosome 12 are the most common genetic abnormalities in patients undergoing therapy.

In the present study, a good agreement was observed between the reduction of viable cells and the induction of cell death in leukemic cells *ex vivo* incubated with the examined agents. Apoptosis induction was reflected by the assessment of such apoptosis indicators as PARP-1 and lamin B cleavage, as well as DNA fragmentation. These events were accompanied by procaspase-9 activation, down-regulation of Mcl-1 and Bcl-2, up-regulation of Bax and Bak, and translocation of histone H1.2 from nuclei to cytoplasm, especially following cell exposure to CM and FM. Recently published reports have revealed that translocation of the H1.2 histone subtype between subcellular compartments is a key event for apoptosis regulation. Apart from the release of proteins from mitochondria e.g. cytochrome c, Smac/DIABLO (second mitochondrial activator of caspase/direct IAP binding protein with low pI) to cytosol, some nuclear proteins (histone H1.2, Nur77) have been described to move in the other direction, i.e., from the nucleus to the cytosol or mitochondria [16, 35, 52]. To date, this observation has been mainly reported as an *ex vivo* event. The mechanism whereby histone H1.2 might affect apoptotic signaling is still unknown. It was suggested that this histone H1 subtype might activate the pro-apoptotic Bak protein by the disrupting the outer mitochondrial membrane and by regulation of apoptosome formation [35]. Our results show that apoptosis induced in leukemic cells *via* CC and FC is associated with the participation of the Bak protein. *Ex vivo* evaluation of cytotoxicity provoked by tested purine analogs combined with an alkylator, as well as analysis of the sub-G1 fraction, DNA ladder formation and apoptosis-related protein expression in model cells exposed to the examined drugs, indicated their high pro-apoptotic potential.

As mentioned in the Results section, the low level of spontaneous apoptosis, confirmed by the Vybrant Apoptosis Assay #4, and the appearance of cleavage products of Mcl-1, procaspase-9, lamin B and PARP-1, was observed in control cells incubated in the culture medium without any drugs (Figs. 1 and 5). The basal expression of anti-apoptotic proteins and the level of spontaneous apoptosis observed in CLL cells incubated exclusively in the culture medium may reflect the general susceptibility of these cells to the treatment. Moreover, the results of the treatment can be manifested to a different degree, depending on some intrinsic properties of CLL cells and their individual responses to the drug(s).

Similarly, high standard deviation values obtained for *in vivo* analyses reflect the wide diversity of CLL cell susceptibility to both applied therapeutic regimens. These results also suggest that individuall patients may demonstrate variable clinical responses to the therapy. It is noteworthy that the observed low level of sub-G1 cells *in vivo* may be caused by rapid engulfment of apoptotic cells by macrophages and some adjacent cells existing in the body – the phenomenon which does not occur *ex* vivo. It has been suggested that the cells undergoing extensive death under *ex vivo* conditions (in the absence of phagocytes) change their phenotype towards necrosis [12]. Moreover, it is well known that apoptotic DNA fragmentation assumes two forms resulting in products that differ in their molecular weight [52, 56]. It has been established that fragmentation of DNA to high molecular weight fragments staying inside the cells after their fixation is undetectable on DNA content histograms [26].

Summarizing, our findings demonstrate the need of *ex vivo* CLL cell treatment before the application of drugs to select the most efficient manner of the patient's therapy. The obtained results confirm the link between the outcomes of the research done under *ex vivo* and *in vivo* conditions and underline the usefulness of *ex vivo* studies in the individual choice of CLL treatment. Our observations provide a base for further studies on the relationship between the *in vivo* clinical responses of patients and *ex vivo* pro-apoptotic activity, and the cytotoxicity of drugs against leukemic cells. Their validation by a study comprising a much larger group of patients is needed.

Acknowledgments:

This work was partially supported by grants No. 505/0375 and No. 545/229 from the University of Łódź. A. Borowiak is a recipient of D-RIM (first edition) fellowship co-funded by the European Union and the European Social Fund, POKL "Human – Best Investment".

The inserted experimental data were firstly published in *Pharmacological Reports* 2013, 65, 460-475.

References:

1   Abramenko IV, Bilous NI, Pleskach GV, Chumak AA, Kryachok IA, Martina ZV, Dyagil IS: CD38 gene polymorphism and risk of chronic lymphocytic leukemia. Leukemia Research 2012, 36, 1237-1240.

2. Alvi AJ, Austen B, Weston VJ, Fegan C, MacCallum D, Gianella-Barradori A, Lane PJ,Hubanek M, Powell JE, Wei W, Taylor MR, Moss PAH, Stankovic T: A novel CDK inhibitor, CYC202 (R-roscovitine), overcomes the defect in p53-dependent apoptosis in B-CLL by down-regulation of genes involved in transcription regulation and survival. Blood, 2005, 105, 4484–4491.

3. Bellosillo B, Vilamor N, Colomer D, Pons G, Montserrat E, Gil J: In vitro evaluation of fludarabine in combination with cyclophosphamide and/or mitoxantrone in B-cell chronic lymphocytic leukemia. Blood, 1999, 94, 2836–2843.

4. Bischoff PL, Holl V, Coelho D, Dufour P, Luu B, Weltin D: Apoptosis at the inter-face of immunosuppressive and anticancer activities: the example of two classes of chemical inducers, oxysterols and alkylating agents. Curr Med Chem, 2000, 7, 693–713.

5. Blobel G, Potter VR: Nuclei from rat liver: isolation method that combines purity with high yield. Science, 1966, 154, 1662–1665.

6. Buggins AGS, Pepper CJ: The role of Bcl-2 family proteins in chronic lymphocytic leukemia. Leuk Res, 2010, 34, 837-842.

7. Castejón R, Vargas JA, Briz M, Berrocal E, Romero Y, Gea-Banacloche JC, Fernández MN, Durantez A: Induction of apoptosis by 2-chlorodeoxyadenosine in B cell chronic lymphocytic leukemia.. Leukemia, 1997, 11, 1253-1257.

8. Chanan-Khan A, Porter CW: Immunomodulating drugs for chronic lymphocytic leukaemia. Lancet Oncol, 2006, 7, 480–488.

9. Cheson BD, Bennett JM, Grever M, Kay N, Keating MJ, O'Brien S, Rai KR: National Cancer Institute – sponsored Working Group guidelines for chronic lymphocytic leukemia: revised guidelines for diagnosis and treatment. Blood, 1996, 46, 4990–4997.

10. Droin NM, Green DR: Role of Bcl-2 family members in immunity and disease. Biochim Biophys Acta, 2004, 1644, 179–188.

11. Eichhorst B, Dreyling M, Robak T, Montserrat E, Hallek M: Chronic lymphocytic leukemia: ESMO Clinical Practice Guidelines for diagnosis, treatment and follow-up. Ann Oncol, 2011, 22, 50-54.

12. Elmore S:. Apoptosis: a review of programmed cell death. Toxicol Pathol, 2007, 35, 495–516.

13. Franiak-Pietryga I, Sałagacka A, Maciejewski H, Błoński JZ, Borowiec M, Mirowski M, Robak T, Korycka-Wołowiec A: Apoptotic gene expression under influence of fludarabine and cladribine in chronic lymphocytic leukemia-microarray study. Pharmacol Rep, 2012, 64, 412-420.

14. Gandhi V, Plunkett W: Combination Strategies for Purine Nucleoside Analogs. in *Chronic Lymphoid Leukemias* ed. Cheson BD, Marcel Dekker, Inc , New York, Basel, 2001 pp. 195-208.

15. Genini D, Adachi S, Chao Q, Rose DW, Carrera CJ, Cottam HB, Carrson DA, Leoni LM: Deoxyadenosine analogs induce programmed cell death in chronic lymphocytic leukemia cells by damaging the DNA and by directly affecting the mitochondria. Blood, 2000, 96, 3537-3543.

16. Giné E, Crespo M, Muntañola A, Calpe E, Baptista MJ, Villamor N, Montserrat, E, Bosch F: Induction of histone H1.2 cytosolic release in chronic lymphocytic leukemia cells after genotoxic and non-genotoxic treatment. Haematologica, 2008, 93, 75-82.

17. Goldin LR, Björkholm M, Kristinsson SY, Turesson I, Landgren O: Elevated risk of chronic lymphocytic leukemia and other indolent non-Hodgkin's lymphomas among relatives of patients with chronic lymphocytic leukemia. Haematologica, 2009, 94, 647-653.

18. Goldstein M, Ross WP, Kaina B: Apoptotic death induced by the cyclophosphamide analogue mafosfamide in human lymphoblastoid cells: Contribution of DNA replication, transcription inhibition and Chk/p53 signaling. Toxicol Appl Pharmacol, 2008, 229, 20-32.

19. Gong J, Traganos F, Darzynkiewicz Z: A selective procedure for DNA extraction from apoptotic cells applicable for gel electrophoresis and flow cytometry. Anal Biochem, 1994, 218, 314–319.

20. Grzybowska-Izydorczyk O, Cebula B, Robak T, Smolewski P: Expression and prognostic significance of the inhibitor of apoptosis protein (IAP) family and its antagonists in chronic lymphocytic leukaemia. Eur J Cancer, 2010, 46, 800-810.

21. Hill MM, Adrian C, Martin SJ: The Mitochondrial Apoptosome Emerges From the Shadows. Mol Cell Biol Lab, 2003, 3, 19-26.

22. Hotz MA, Gong J, Traganos F, Darzynkiewicz Z: Flow cytometric detection of apoptosis: comparison of the assays of in situ DNA degradation and chromatin changes. Cytometry, 1994, 15, 237–244.

23. Igney FM, Krammer PH: Death and anti-death: tumor resistance to apoptosis. Nat Rev Cancer, 2002, 2, 277–288.

24. Jain P, O'Brien S: Richter's Transformation in Chronic Lymphocytic Leukemia. Oncology, 2012, 26, 1- 12.

25. Johnston JB, Paul JT, Neufeld NJ, Haney N, Kropp DM, Hu X, Cheang M, Gibson SB: Role of myeloid cell factor-1 (Mcl-1) in chronic lymphocytic leukemia. Leuk Lymph, 2004, 45, 2017–2027.

26. Kajstura M, Halicka HD, Pryjma J, Darzynkiewicz Z: Discontinuous fragmentation of nuclear DNA during apoptosis revealed by discrete „sub-G_1" peaks on DNA content histograms. Cytometry, 2007, 71, 125–131.

27. Keating MJ, O'Brien S, Kontoyiannis D, Plunkett W, Koller C, Beran M, Lerner S, Kantarjian H: Results of the first salvage therapy for patients refractory to a fludarabine regimen in chronic lymphocytic leukemia. Leuk Lymph ,2002, 43, 1755-1762.

28. Kitada S, Andersen J, Akar S, Zapata JM, Takayama S, Krajewski S, Wang H-G, Zhang X, Bullrich F, Croce CM, Rai K, Hines J, Reed JC.: Expression of apoptosis-regulating proteins in chronic lymphocytic leukemia: correlation with in vitro and in vivo chemoresponses. Blood, 1998, 91, 3379–3389.

29. Klein U, Dalla-Favera R: Germina centers: role in B-cell physiology and malignancy. Nature Rev Immunol, 2008, 8, 22-33.

30. Kobylińska A, Błoński JZ, Hanausek M, Wałaszek Z, Robak T, Kiliańska ZM: Determination of the in vivo effects of cladribine alone and its combinations with cyclophosphamide or cyclophosphamide and mitoxantrone on Bax and Bcl-2 protein expression in B-CLL cells. Oncol Rep, 2004, 11, 699–705.

31. Kobylińska A, Bednarek J, Błoński JZ, Hanausek M, Wałaszek Z, Robak T, Kiliańska ZM: In vitro sensitivity of B-cell chronic

lymphocytic leukemia to cladribine and its combination with mafosfamide and/or mitoxantrone. Oncol Rep, 2006, 16, 1389–1395.

32. Kristinsson SY, Dickman PW, Wilson WH, Caporaso N, Björkholm M, Landgren O: Improved survival in chronic lymphocytic leukemia in past decade: a population based study including 11179 patients diagnosed between 1973-2003 in Sweden. Haematologica, 2009, 94, 1259–1265.

33. Laemmli UK: Cleavage of structural proteins during the assembly of the head of bacteriophag T4. Nature, 1970, 227, 680–685.

34. Leary JJ, Brigati JJ, Ward DC: Rapid and sensitive colorimetric method for visualizing biotinlabeled DNA probes hybridized to DNA or RNA immobilized on nitrocellulose: Bioblots. Proc Natl Acad Sci USA, 1983, 80, 4045–4049.

35. Lindenboim L, Borner C, Stein R: Nuclear proteins acting on mitochondria. Biochim Biophys Acta, 2011, 1813, 584-596.

36. Lowry OH, Rosebrough NJ, Farr AL, Randall RJ: Protein measurement with the Folin phenol reagent. J Biol Chem, 1951, 193, 265–275.

37. Maddocks KJ, Lin TS: Update in the management of chronic lymphocytic leukemia. J Hematol Oncol, 2009, 2, 29 doi; 10.1186/1756-8722-2-29.

38. Marzo I, Perez-Galan P, Giraldo P, Rubio-Felix D, Anel A, Naval J: Cladribine induces apoptosis in human leukaemia cells by caspase-dependent and -independent pathways acting on mitochondria. Biochem J, 2001, 35, 537–546.

39. Mazur L, Opydo-Chanek M, Stojak M, Wojcieszek K: Mafosfamide as a new anticancer agent: preclinical investigations and clinical trials. Anticancer Res, 2012, 32, 2783-2789.

40. Messmer BT, Messmer D, Allen SL, Kolitz JE, Kudalkar P, Cesar D, Murphy EJ et al.: In vivo measurements document the dynamic cellular knetics of chronic lymphocytic leukemia B cells. J Clin Invest, 2005, 115, 755–764.

41. Michels J, Johnson PW, Packham G: Mcl-1. Int J Biochem Cell Biol, 2005, 37, 267–271.

42. Nicoletti I, Migliorati G, Pagliacci MC, Grignani F, Riccardi C: A rapid and simple method for measuring thymocyte apoptosis by propidium iodide staining and flow cytometry. J Immunol Methods, 1991, 139, 271–279.

43. O'Brien SM, Kantarjian HM, Cortes J, Beran M, Koller CA, Giles FJ, Lerner S, Keating M: Results of the fludarabine and cyclophosphamide combination regimen in chronic lymphocytic leukemia. J Clin Oncol, 2001, 19, 1414–1420.

44. Pepper C, Lin TT, Pratt G, Hewamana S, Brennan P, Hiller L, Hills R et al.: Mcl-1 expression has in vitro and in vivo significance in chronic lymphocytic leukemia and is associated with other poor prognostic markers. Blood, 2008, 112, 3538–3540.

45. Rai KR, Sawitsky A, Cronkite EP, Chanana AD, Levy R, Pasternack BS: Clinical staging of chronic lymphocytic leukemia. Blood, 1975, 46, 219–234.

46. Reed JC, Pellecchia M: Apoptosis-based therapies for hematologic malignancies. Blood, 2005, 106, 408–418.

47. Reed JC: Bcl-2 family proteins and hematologic malignancies: history and future prospects. Blood, 2008, 111, 3322–3329.

48. Robak T, Błoński JZ, Kasznicki M, Góra-Tybor J, Dwilewicz-Trojaczek J, Stella-Hołowiecka B, Wołowiec D: Cladribine combined with cyclophosphamide is highly effective in the treatment of chronic lymphocytic leukemia. Hematol J, 2002, 3, 244–250.

49. Robak T, Błoński JZ, Góra-Tybor J, Jamroziak K, Dwilewicz-Trojaczek J, Tomaszewska A, Konopka L et al.: Cladribine alone and in combination with cyclophosphamide or cyclophosphamide plus mitoxantrone in the treatment of progressive chronic lymphocytic leukemia: report of a prospective, multicancer, randomized trial of the Polish Adult Leukemia Group (PALG CLL2). Blood, 2006, 108, 473–479.

50. Robak T, Błoński JZ, Wawrzyniak E, Góra-Tybor J, Palacz A, Dmoszyńska A, Konopka L, Warzocha K, Jamroziak K: Activity of cladribine combined with cyclophosphamide in frontline therapy for chronic lymphocytic leukemia with 17p13.1/*TP53* deletion. Cancer, 2009, 115, 94–100.

51. Robak T, Jamroziak K, Góra-Tybor J, Stella-Hołowiecka B, Konopka L, Ceglarek B, Warzocha K et al.: Comparison of cladribine plus cyclophosphamide with fludarabine plus cyclophosphamide as first-line therapy for chronic lymphocytic leukemia: a phase III randomized study by the Polish Adult Leukemia Group (PALG-CLL3 Study). J Clin Oncol, 2010, 28, 1863–1869.

52. Robertson JD, Orrenius S, Zhivotovsky B: Nuclear events in apoptosis. J Struct Biol, 2000, 129, 346-58

53. Rogalińska M, Błoński JZ, Hanausek M, Wałaszek Z, Robak T, Kiliańska ZM: 2-Chlorodeoxyadenosine alone and in combination with cyclophosphamide and mitoxantrone induce apoptosis in B chronic lymphocytic leukemia cells in vivo. Cancer Detect Prev, 2004, 28, 433–442.

54. Rogalińska M, Góralski P, Woźniak K, Bednarek JD, Błoński JZ, Robak T, Piekarski H, Hanausek M, Walaszek Z, Kilianska ZM:; Colorimetric study as a test for choosing treatment of B-cell chronic lymphocytic leukemia. Leuk Res, 2009, 33, 308–314.

55. Rogalińska M, Błoński JZ, Komina O, Góralski P, Żołnierczyk JD, Piekarski H, Robak T, Kiliańska ZM, Węsierska-Gądek J: R-roscovitine (Seliciclib) affects CLL cells more strongly than combinations of fludarabine or cladribine with cyclophosphamide: Inhibition of CDK7 sensitizes leukemic cells to caspase-dependent apoptosis. J Cell Biochem, 2010, 109, 217–235.

56. Samejima K, Tone S, Earnshaw WC:. CAD/DFF40 nuclease is dispensable for high molecular weight DNA cleavage and stage I chromatin condensation in apoptosis. J Biol Chem, 2001, 276, 45427–45432.

57. Saxena A, Viswanathan S, Moshynska O, Tandon P, Sankaran K, Sheridan DP:. Mcl-1 and Bcl-2/Bax ratio are associated with treatment response but not with Rai stage in B-cell chronic lymphocytic leukemia. Am J Hematol, 2004, 75, 22–33.

58. Sieklucka M, Pożarowski P, Bojarska-Junak A, Hus I, Dmoszyńska A, Roliński J: Apoptosis in B-CLL: The relationship between higher ex vivo spontanous apoptosis before treatment in III-IV Rai stage patients and poor outcome. Oncol Rep, 2008, 19, 1611-1620.

59. Stevenson FK, Caligarias-Cappio F: Chronic lymphocytic leukemia: revelations from the B-cell receptor. Blood, 2004, 103, 4389-4395.

60. Towbin H, Staechelin T, Gordon J: Electrophoretic transfer of protein from polyacrylamide gels to nitrocellulose sheets: procedure and some applications. Proc. Nat Acad Sci USA, 1979, 76, 4350–4354.

61. Umansky SR, Korol BR, Nelipovich PA: In vivo DNA degradation in the thymocytes of γ-irradiated or hydrocortisone-treated rats. Biochim Biophys Acta, 1981, 655, 281–284.

62. Van den Neste E, Cardeon S, Offner F, Bontemps F: Old and new insights into the mechanisms of action of two nucleoside analogs active in lymphoid malignancies: fludarabine and cladribine. Int J Oncol, 2005, 27, 1113-1124.

63. Wieder T, Essmann F, Prokop AK, Schulze-Osthoff K, Beyaert R, Dorken B, Daniel PT: Activation of caspase-8 in drug-induced apoptosis of B-lymphoid cells is independent of CD95/Fas receptor-ligand interaction and occures downstream of caspase-3. Blood, 2001, 97, 1378-1387.

64. Willimot S, Merriam T, Wagner SD: Apoptosis induces Bcl-X$_S$ and cleaved Bcl-X$_L$ in chronic lymphocytic leukemia. Biochem Biophys Res Commun, 2011, 404, 480–485.

65. Zenz T, Frohling S, Martens D, Dohner H, Stilgenbauer S: Moving from prognosis to predicative factors in chronic lymphocytic leukemia (CLL) . Best Pract Res Clin Hematol, 2010, 23, 71-84.

66. Żołnierczyk JD, Błoński JZ, Robak T, Kiliańska ZM, Węsierska-Gadek J: Roscovitine triggers apoptosis in B-cell chronic lymphocytic leukemia cells with similar efficiency as combinations of conventional

purine analogs with cyclophosphamide. Ann NY Acad Sci, 2009, 1171, 124–131.